Acknowledgements

Introduction

This booklet is for subject coordinators and teachers in primary schools. It has been produced in response to requests from many schools for guidance on standards in the non-core subjects at Key Stages 1 and 2. As there are no plans to extend the requirements for statutory teacher assessment to the non-core subjects in these key stages, the emphasis here is on exemplifying standards to support planning, teaching and assessment.

Illustrating expectations

To help in planning, this booklet looks at standards by focusing on what most children are expected to know, understand and be able to do in geography by the end of Year 2, Year 4 and Year 6. It is recognised that not all children will reach these expectations and some children will exceed them. These expectations are illustrated using examples of children's work.

Using the programmes of study and level descriptions to establish expectations

In identifying these expectations this booklet draws on the geography programmes of study for Key Stages 1 and 2 and on the attainment outlined in the level descriptions. In geography these expectations for children by the end of Years 2, 4 and 6 relate to Levels 2, 3 and 4 respectively.

Planning for curriculum continuity and children's progress

A sound understanding of the expectations in geography will help to inform your planning, teaching and assessment. The booklets in this series complement *Planning the Curriculum at Key Stages 1 and 2* (SCAA, 1995) by helping staff establish whole-school expectations for children's learning in the non-core subjects, recognising that each is part of the broader primary curriculum. This will help you plan for progress in children's learning and for continuity in children's experience across Key Stages 1 and 2.

Using this booklet

This booklet is in four sections.

Section 1

Section 1 sets out the expectations for children's attainment by the end of Years 2, 4 and 6 and shows how children are likely to progress during Key Stages 1 and 2. It also indicates how the expectations reflect the programmes of study and level descriptions in the geography Order.

Sections 2, 3 and 4

Sections 2, 3 and 4 illustrate these expectations using work from a number of children. For each example of work, the introductory text explains the context for the work, and the references to the programme of study highlight the points which the teacher planned to address. The comment in the margin indicates how the work illustrates the expectations.

Planning units of work and assessment

These expectatons can be used:

- in long term planning, to establish whole-school expectations for children's work when devising units of work in geography, either on their own or in combination with other subjects;

- in medium and short term planning, as a basis for developing learning objectives matched to the abilities of the children in each class. You may also find it helpful to refer to *Teacher Assessment in Key Stage 2* (SCAA, 1997) which looks at the development of learning objectives;

Developing a school portfolio

- as a basis for developing a school portfolio, which can act as a reference document for the standards agreed within the school. Many schools have found school portfolios in geography useful both in planning work and in making assessments. You may find it helpful to refer to *Consistency in Teacher Assessment: Guidance for Schools* (SCAA, 1995);

Reporting to parents

- as a device for comparing the performance of individual children against typical expectations. This can help in writing the annual report to parents on each child's progress in National Curriculum subjects.

Further advice on planning the geography curriculum at Key Stage 2 is available in *Geography at Key Stage 2: Curriculum Planning Guidance for Teachers* (SCAA, 1997).

Section 1: Expectations in geography

Expectations and the geography Order

This section identifies the main aspects in which children demonstrate attainment and make progress in geography. It also clarifies how the expectations reflect the programmes of study for Key Stages 1 and 2 and the level descriptions in the geography Order.

Programmes of study

The programmes of study for Key Stages 1 and 2 set out the content to be taught under the heading of skills (paragraphs 2 and 3), places (paragraphs 4 and 5) and themes (paragraph 6 onwards). An enquiring, questioning approach to geographical work is also required (paragraphs 1b and 2).

Level descriptions

The level descriptions add to this information in two ways. The range of places and themes studied and the scale of study are referred to in broad bands of levels (ie 1 to 3; 4 and 5; 6 to 8; Exceptional). These provide the context within which children demonstrate knowledge, understanding of geographical ideas and competence in geographical enquiry. They are dependent on the programme of study content and on the range of additional topics which you choose to provide. For instance, for Levels 4 and 5, children are required to 'show their knowledge, understanding and skills in relation to studies of a range of places and themes, at more than one scale.' The level descriptions also highlight the type and range of performance likely to be demonstrated by children.

Aspects where progress and expectation are shown

When the level descriptions are considered alongside the programme of study content, it is possible to identify four aspects of geography in which progress and expectation are shown. These are:

- knowledge and understanding of **places**;

- knowledge and understanding of **patterns and processes**;*

- knowledge and understanding of **environmental relationships and issues**;

- ability to undertake **geographical enquiry** and to use **skills**.

*Note: In geography, the way that physical features and human features are arranged in a landscape or occur in an environment is referred to as a **pattern** (eg the layout of hedgerows in a farming landscape; the trends in temperature change through the seasons; the way that streets are arranged in a town). The term **process** is used in geography for a series of events which cause change (eg the wearing away of a bank by the flow of the river changes the shape of the channel; the residential redevelopment of old warehouses in a town changes the character and use of that area).

Broad lines of progression can be recognised for each of these aspects running through the level descriptions from Level 1 to Exceptional Performance. In the table which follows, the main expectations for performance in each of these four aspects have been summarised for the end of Year 2 (Level 2 in the context of the Key Stage 1 Programme of Study), the end of Year 4 (Level 3 in the context of the Key Stage 2 Programme of Study) and the end of Year 6 (Level 4 in the context of the Key Stage 2 Programme of Study).

ASPECTS	By the end of Year 2
PLACES Features and character of places	Describe the main features of localities they have studied, using appropriate geographical terms (eg hill, river, factory) and demonstrate (eg orally, in pictures, in writing) that they recognise those features that give localities their character (eg busy shopping street, large flat fields).
Contrasts and relationships between places	Begin to recognise contrasts in individual features of different localities (eg characteristics of houses in photographs of a tropical locality and their own locality).
PATTERNS AND PROCESSES Patterns	Respond to questions about 'where things are', by making simple observations about features in the environment (eg the shop is in the village centre, the pedestrian crossing is next to the school, summer is warmer than spring).
Physical and human processes	Respond to questions about 'why things are like that', by recognising and making appropriate observations about physical features (eg the heavy rain has flooded the school field) and human features (eg the village shop has closed because not enough people use it).
ENVIRONMENTAL RELATIONSHIPS AND ISSUES Viewpoints and perspectives about the environment	Express their own views about the physical and/or human features of the environment in a locality (eg a wild rocky landscape, a noisy street).
Environmental interactions and management	Recognise how an environment changes and how people are affecting an environment, when asked 'what is this place like?' (eg seasonal change in the landscape, rubbish tipped in a local beauty spot).
GEOGRAPHICAL ENQUIRY AND SKILLS Geographical enquiry	Ask and respond to questions about places and topics studied, on the basis of information provided by the teacher and their own observations (eg recognise the main landscape features on a photograph).
Use of skills	Undertake simple tasks, using maps, diagrams and other secondary sources as demonstrated by the teacher (eg use letter and number co-ordinates to identify a feature, draw a pictorial diagram of shops in the High Street).

By the end of Year 4	By the end of Year 6
Describe a range of physical and human features of localities studied, using appropriate geographical terms (eg transport, industry, tributary) and begin to offer reasons for the distinctive character of different places (eg the coastal location and attractive scenery of a seaside resort).	Describe the physical and human features of a range of places studied and show how the mix of these features helps to explain their character (eg explaining how the weather, landscape and human activities in a hill farming area make it very different from an urban locality).
Make geographical comparisons between the localities studied and begin to offer reasons for their findings, including reference to location (eg referring to situation and weather when comparing the scenery and activities of a tropical locality with the home locality).	Draw out similarities and differences between places, including awareness of their wider geographical location (eg which region; which country; which continent?) and understanding of links between them (eg trade between countries).
Offer appropriate observations about locations and the patterns made by individual physical and human features in the environment (eg hotels along the sea-front, frost on the northern side of the school playground).	Begin to appreciate the importance of location in understanding places and offer explanations for patterns of physical or human features (eg explain why a town grew up at a river crossing).
Begin to explain 'why things are like that', referring to physical and human features of the landscape (eg why factories near motorways can receive goods and materials easily, how exposure to sun or shade can affect snow conditions in ski resorts).	Recognise selected physical and human processes (eg river erosion, closure of a coal mine) and begin to appreciate how these can change the character of places and environments they have studied.
Begin to account for their own views about the environment, recognising that other people may have reasons for thinking differently (eg explaining why a rocky coastline is attractive to them, but may be dangerous for fishermen).	Identify and explain the different views held by people about an environmental change (eg the different views held by residents and the developer about plans to build a fast food restaurant).
Identify how people affect the environment and recognise ways in which people try to manage it for the better (eg recognise that restricting car access is one way to reduce air pollution in the local town).	Recognise and describe how people can improve or damage the environment in particular cases, and describe different approaches taken to management (eg investigate alternative strategies for minimising erosion of a footpath).
Ask and respond to geographical questions (eg why are there fewer trees on the hill top?) in the course of undertaking tasks set by the teacher, and offer their own ideas appropriate to the situation (eg suggestions for carrying out a fieldwork task).	Draw on their own observations and on secondary sources provided, and use their awareness of topical matters to suggest geographical questions and issues which might be studied (eg impact of a hurricane, effect of a new local superstore).
Use a range of simple pieces of equipment and secondary sources (eg atlas, photographs, anemometer) to carry out tasks supported by the teacher (eg find the correct page in the atlas, measure the wind speed).	Use confidently a full range of skills (specified for Key Stage 2) and different kinds of maps and resources, to undertake some independent investigations and some planned by the teacher (eg take measurements of river speed on fieldwork, record the results using IT, and plot them on a diagram).

Section 2: Expectations by the end of Year 2

The following examples, taken from the work of a number of children, illustrate the expectations in geography by the end of Year 2.

What and where are the countries of the UK?

Programme of study:
1b, 1c, 2, 3e, 4

The teacher wanted children to be able to name the constituent countries of the UK and the province of Northern Ireland. The stimulus was provided by the story of *The Lighthouse Keeper's Lunch*. The children used maps, floor puzzles and atlases to discover where lighthouses might be found and why they might be found there. Afterwards, a simple, computer-based exercise was used to test their locational knowledge. They had to position correctly prepared country names on a map of the UK. Then they were asked to reassemble a computer jigsaw map of the UK, to make their own labels and put them on the correct countries. The teacher followed this activity by asking the children to identify and describe the location of their own locality within the UK.

Amy and Josh have responded to the questions asked about the UK and carried out both simple map and computer-based tasks. They recognise the countries of the UK and label the first map correctly. In the second activity, they complete the jigsaw and locate all their own labels appropriately, apart from the Northern Ireland one.

(Geographical Enquiry and Skills, Patterns)

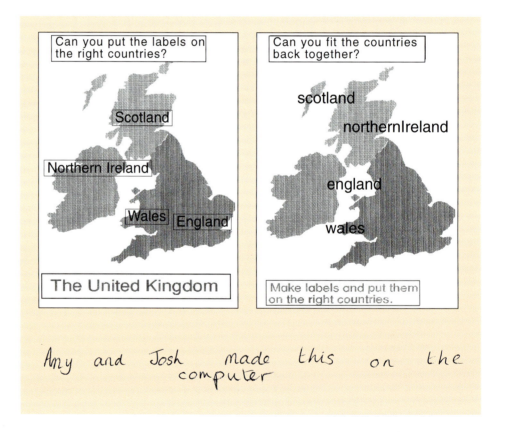

Amy and Josh made this on the computer

8

Near and far

Programme of study:
1a, 1c, 3a, 3d, 3f, 4

The teacher wanted to find out to what extent children were aware of the world beyond their locality. She asked the children to look at some photographs of different places (the local park, a nearby town, London, a village in India, high-rise office blocks in Hong Kong) and, in groups, to sort them in terms of their distance from home (near and far). After class discussion, the children were each given a blank near/far diagram on which to draw their own examples for each category.

Edward responds creatively to the task. He includes in his diagram a wide range of places, based on his own experience and going beyond the photographs shown in class. He shows that he can distinguish different grades of near and far (relative location).

(Geographical Enquiry and Skills, Places, Patterns)

Teacher: *Edward, why did you think France was far away from your home but that India was very far away?*

Edward: *Well, when we went to France, we got on a ferry, but I think you'd have to get a plane for India.*

The dialogue reveals some of his reasoning and an understanding of the links between places (eg ferry travel to France, air travel to India). Further work on children's holidays and means of travel might be one way to develop this understanding.

(Places)

Programme of study:
1b, 2, 3b, 3c, 3d, 4, 5a, 6

Getting to know the school grounds

The children walked around the school grounds, using a large scale plan to identify buildings, playground, grass, the pond and the nursery. The teacher wanted to introduce them to following directions and using maps as they explored their immediate locality.

Back in the classroom, the children colour-coded a copy of the plan. When asked what they liked and disliked about the school grounds, the children drew pictures, labelled them and talked about them. These activities were developed further during the visit to the local river, described on page 11.

Amber draws on her fieldwork experience to identify the main features of the school grounds. She colour-codes the plan correctly, although she does not completely understand the difference between grid lines and boundaries on the map.

(Geographical Enquiry and Skills, Places)

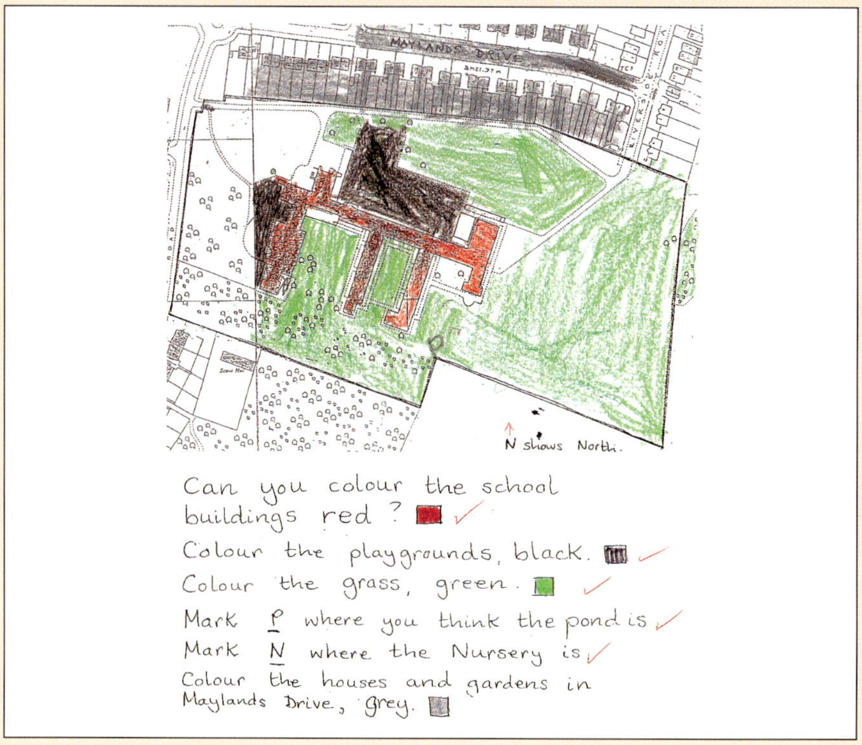

N shows North.

Can you colour the school buildings red? ✓
Colour the playgrounds, black. ✓
Colour the grass, green. ✓
Mark P where you think the pond is ✓
Mark N where the Nursery is ✓
Colour the houses and gardens in Maylands Drive, grey.

She expresses her own views clearly in the pictorial diagram and, in conversation with the teacher, is able to explain some of her reasons.

(Environmental Relationships and Issues)

I Like This park
I dislike the
Smells nice
Royal Park School
pretty Rose
Noisy place
quiet place

Amber:
I don't like the toilets 'cos they are smelly. That's nice by the pond – you can be quiet all by yourself.

Programme of study:
1a, 1b, 2, 3a, 3b, 6

What is the local river like?

The teacher wanted the children to recognise the physical and human features of a place. A field visit was planned to the local river (a site known to many of them). In preparation, the children were asked to draw pictures showing what they would expect to find there. They then made suggestions for a questionnaire, which was completed during the field visit.

Amber's picture of the River Cray at 'Five Arches', prepared before the field visit, shows that she has responded to the question in an appropriate way. She recognises and describes pictorially the main features which give this place its character. She also shows some recognition of how people affect the environment (pollution).

(Places, Geographical Enquiry and Skills, Environmental Relationships and Issues)

Amber completes this simple questionnaire effectively, and her response confirms that she recognises appropriate features.

(Geographical Enquiry and Skills)

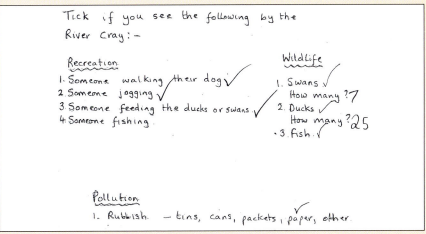

Extract from longer piece of work

What are hot deserts like?

As part of a topic on deserts, linked with work in science and English, the children watched a film about the hot desert environment in the countries of the Sahara region. The teacher wanted them to understand how plants adapt to the environmental conditions. In groups they carried out an experiment to find out how important moisture and temperature were for plant growth (cress seeds). The children were then asked to relate their findings to some of the plants seen in the film. They went on to consider how humans adapt to living in the hot desert environment.

Programme of study:
1b, 1c, 2, 3f, 5a, 5c
[also **Science**: Experimental and Investigative Science: 2, 3 Living Things: 3]

Francesca records the results from the group experiment and her brief conclusion shows that she recognises why plant growth varies. In her written work she draws on the information seen in the film and makes the link back to the hot desert environment. In this respect she shows an awareness of a place beyond her locality and some understanding of why it is like that.

(Geographical Enquiry and Skills, Patterns and Processes, Places)

In contributing to the class display about hot deserts, Francesca responds to the questions posed about this environment and recognises the features characteristic of this place.

(Geographical Enquiry and Skills, Places)

> **Which pot will grow:**
>
> | | Oct 14th | Oct 15th | Oct 17th | Oct 18 |
> |---|---|---|---|---|
> | 1. Room temp. no water | nothing | nothing | nothing | nothing |
> | 2. ✓ Room temp. water when needed | nothing | 2cm high | 5cm | 6cm |
> | 3. Room temp. water only once. | seeds germinating | 2cm high | 4cm | 6cm |
> | 4. Water. Put in fridge. | nothing | nothing | nothing | nothing |
> | 5. Water in till seeds grow. Put in a hot place. | seeds germinate | 5cm high | 6cm drooping | Now drooping badly |
>
> Seeds need water and warmth to germinate. No. 5 grew very well but it did not have enough water so it began to die.

> **Adapting**
>
> (changing)
>
> Animals, plants and people can adapt to living in a desert.
>
> Some grasses in an oases can stretch its roots 50 metres long to get water.
>
> The tamarisk bush can stretch is roots 5 metres deep to also get water.
>
> Some seeds can be buried in the desert untill rain falls then the seeds germinate very quickly.

12

Programme of study:
1a, 1b, 2, 3b, 6a, 6b, 6c

The environment in our local shopping street

A joint group of Year 1 and 2 children were taken out to visit the local shopping street (Norwich Road). They had already studied the neighbourhood shops and talked about some photographs of the Norwich Road area. While Year 1 children were identifying those shops they needed to visit to collect ingredients on a shopping list, Year 2 children were examining what the environment was like and how it might be improved. Their investigative work was structured by means of survey sheets and, although the survey work was carried out in groups, each child wrote up the results independently for a project folder.

Stephen's survey sheet shows that he understands the task and can record traffic correctly. In his comments to his teacher, he reveals that he makes appropriate observations about the place and begins to recognise similarities and differences, *eg more traffic at the Rotterdam Road end.*

(Geographical Enquiry and Skills, Places)

What is Norwich Road like?

1. Traffic - survey busy or quiet

type of vehicle	number of vehicles in 10 minutes
cars	Rotterdam Road Mawari Street
vans	4
lorries	6
bicycles	3
motor bikes	
others	0 8

Stephen: (to teacher)
We found out that there was more traffic at the Rotterdam Road end, even though there were less shops.

Stephen's pictures and comments to the teacher express his views about the environment and his reasons for these.

(Environmental Relationships and Issues)

Stephen:
The metal bits sticking out of the building (the cinema) are dangerous – a double decker bus could hit them.

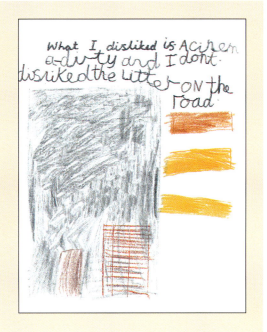

What I disliked is A citchen a dirty and I dont. disliked the litter on the road

Section 3: Expectations by the end of Year 4

The following examples, taken from the work of a number of children, illustrate the expectations in geography by the end of Year 4.

Jobs at the harbour

As part of a combined geography and design & technology unit of work, Year 4 children visited the local harbour. The teacher wanted the children to appreciate how jobs often reflect the location and character of the settlement. The harbour master explained some aspects of work in the harbour and the children were shown around the different buildings. They recorded their findings in notes and sketches. Back at school, they were asked to illustrate six different jobs which they had seen being done, to identify on a large scale map those places where they had seen the work taking place, and then to mark these locations on their worksheet sketch map. A class discussion highlighted the reasons why certain jobs were located at the harbour and some children were able to explain this in their writing.

Programme of study:
1b, 1d, 3b, 3d, 5c, 9a, 9b

Charlotte illustrates six different jobs she saw during the fieldwork. She also locates these on the sketch map of the harbour.

(Geographical Enquiry and Skills, Places)

Charlotte's more detailed description of the fishermen's huts is evidence that she recognises some of the working activities carried out in a harbour locality, and she has observed and recorded these carefully.

(Geographical Enquiry and Skills, Processes)

Programme of study:
1b, 1c, 3a, 3c, 4, 5a, 5c

Using photographs to investigate a locality

As an introduction to aerial photographs, the children examined A4-size oblique aerial photographs of a contrasting UK locality. In ability groups, they were given an outline plan showing the roads and some key land use boundaries. They were also given lists of features (word bank) and were asked to identify these features from the photograph and label them clearly on the plan.

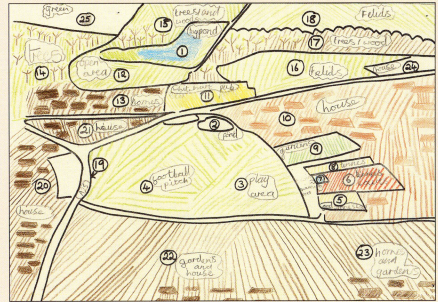

Angela's plan is clearly labelled using appropriate colours and symbols. She recognises features on a photograph and a plan. Apart from a spelling mistake ('felids'), she communicates her findings effectively.

(Geographical Enquiry and Skills, Places)

In the class discussion which follows, she is beginning to recognise patterns and offer explanations for locations.

(Patterns and Processes, Places)

Teacher: *Is there anything special that you have noticed about the village?*
Angela: *The old village is where the roads meet, and the houses have spread along it (the road).*

Comparing localities

The children visited the village of Ilam in Derbyshire to compare another UK locality with their own. The fieldwork was undertaken on a Year 4 residential week, during which time the children observed life in the village, visited a farm and carried out a river investigation. Each child kept a diary of the activities and one of the tasks on return to school was to draw up a table of contrasts between Ilam and Furzton, their home in Bletchley.

> It is a very small village very quiet much different than Bletchley. We found a net under a bridge in the water (River Manifold) Mrs Grimble said that it was a net to slow the water down under the bridge.

	Furzton	Ilam
Surroundings:-	It is busy with lots of parks and not many fields.	It has lots of fields and animals. Trees and & rivers. ✓
Noises./Level.	You can hear and lorrys. buses.	lots of cars Children shouting. Animals, birds rivers glowing.
Busy?	It is very crowded.	busy and It is quiet not at all busy not at all crowded.
Buildings.	They are quiet new they are mostley made out of brick They look	They are oldish the are mostley made of stone they look very coulo colourful. old fasioned
Population	There are 10s 10s people in Furton It is very busy.	There arn't so many people it is very quiet.

Extract from longer piece of work

what we are comparing in....	Furzton	Ilam
Jobs	There are lots of jobs to get because we have the city.	There arn't so many jobs because there arn't so many shops.
communications	There are lots of roads and it isn't very far from the air port.	There are twisty roads and not very many car of buses were used.

Extract from longer piece of work

Sidebar:

Programme of study:
1a, 1b, 1c, 1d, 3a, 3b, 5a, 5b, 9a, 9b

Rachel understands that Ilam is in a different part of England to her home locality and her diary extract reveals that she is aware of some features which give Ilam its character, *eg small village, very quiet, river.*

(Places)

In her table of comparisons, she draws on her fieldwork experience to highlight some of the differences between individual features of the two places, and begins to state reasons, *eg because we have the city.*

(Places, Geographical Enquiry and Skills)

Programme of study:
1b, 2, 3b, 3d, 3e, 10a, 10b

Making plans for the old station site

The children were investigating environmental change using the example of a derelict station site in the local area. The project started with the question, 'What shall we do with the old station site at Redbourn?' and the children brainstormed further enquiry questions to investigate. They then followed a full enquiry sequence of fieldwork, surveys, discussion of alternatives and making recommendations about the site. For part of the time the children worked with a planner from the local planning department, but after analysing the character and layout of the site, the children put forward their own individual plans. Some children made presentations to the class.

Katy explains the enquiry sequence she followed. She recognises that different views might be held about the site and also presents her own viewpoint clearly, *eg But it does matter....*

(Geographical Enquiry and Skills, Environmental Relationships and Issues)

Extracts from longer pieces of work

Her completed site plan shows that she can use and interpret a large scale plan. Her proposals provide evidence that she recognises ways in which this environment could be managed. Katy did not make a presentation to the class, although a written explanation of her plans would have provided further information about her performance.

(Geographical Enquiry and Skills, Environmental Relationships and Issues, Patterns and Processes)

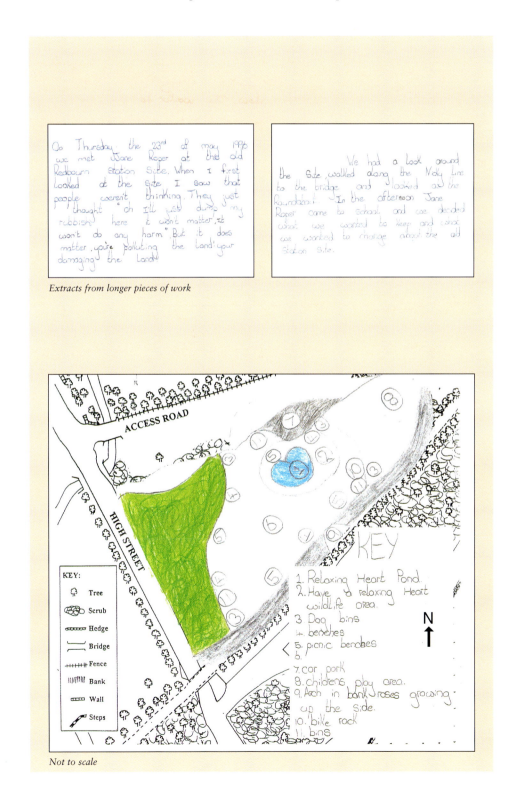

Not to scale

The Arctic Year

Programme of study:
1c, 1d, 3a, 3e, 8b, 8c

The children were learning about weather patterns around the world. The teacher used a globe and a torch to explain how the sun's rays affect the earth and the children then completed worksheets to test their understanding. Finally, using photographs and stories, the teacher explained what the pattern of weather and life is like in the Arctic. Each child was asked to design a poster showing the seasons of the Arctic year.

Natalie's 'Arctic Year' poster is logically arranged and clearly labelled to communicate the information. She shows some understanding of seasonal change, as well as awareness of the character of the place. In her picture and description, Natalie explains why the Arctic is like that, and links climate, wildlife and human activity.

(Geographical Enquiry and Skills, Patterns and Processes, Places)

Programme of study:
1b, 1c, 2, 3a, 3b, 3d, 8a

The weather around our school grounds

A Year 4 class was studying how site conditions can influence weather. They had made predictions about the places in the school grounds which would be the warmest, coldest, most windy and least windy. After demonstrations of how to use the measuring equipment (anemometer, compass, thermometer) small groups of children each visited two out of ten sites identified around the school. With supervision from an adult helper, the children measured and recorded wind speed, direction and temperature. The results were discussed back in class and compared with the predictions. A weather recording activity, which extended children's skills by using IT, is shown on page 23.

According to her teacher, Katy made sensible predictions before the fieldwork and was able to undertake the tasks of finding the sites and measuring with the weather equipment. Her records are neat and accurate.

(Geographical Enquiry and Skills)

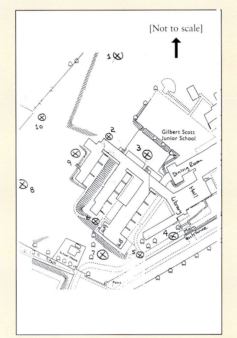

| PLACE | WIND | | TEMPERATURE |
	DIRECTION	SPEED	
6	E	0 MPH	5°C
3	S	1 MPH	8°C
7	E	0 MPH	13°C
4	W	0 MPH	9°C
5	SW	0 MPH	7°C
2	E	2 MPH	10°C
1	W	2 MPH	12°C
9	N	3 MPH	13°C
10	NE	4 MPH	10°C
8	NE	5 MPH	10°C

Teacher comment: *Katy's group was one which showed considerable independence. Her group made its own way to the two sites and used their measuring instruments with little help. Katy's report was her own work and she had managed to predict where it would be warmest and coldest, based on her personal experience.*

In her summary Katy identifies similarities and differences between the sites and begins to give reasons for these, *eg because of the school buildings.*

(Patterns and Processes)

> The Weather around Our School
> We went to places around our school. There were ten different places. We found out that the warmest places were warm because they had the sun on them nearly all day. The coolest places were cool because they were in shade. Then we found out that the windiest places were windy because they had no shelter and least windy was because it was in shelter. Most things are in shade or shelter because of the school buildings or trees.

Section 4: Expectations by the end of Year 6

The following examples, taken from the work of a number of children, illustrate the expectations in geography by the end of Year 6.

Using the 1:50,000 map to investigate settlement

Having talked in class about the sites of settlements and studied pictures and sketch maps, Year 6 children were each given a 1:50,000 map extract of an area in England. The teacher wanted the children to develop their skills of using the 1:50,000 map for selecting information. They were asked to find examples of settlements that had grown up on the following sites: river bend, natural harbour, major junction, near a castle, on a hill, on the side of a valley. A sketch map and a commentary were required for each example.

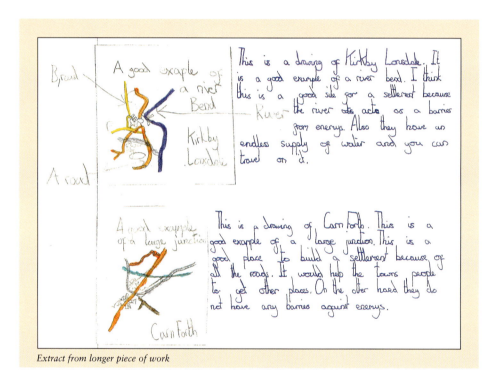

Extract from longer piece of work

Programme of study:
1a, 1c, 1d, 2, 3a, 3d, 9a

Jack's sketch maps show that he can select relevant information from the OS 1:50,000 using the symbols and contour lines. In his commentaries he shows understanding of the importance of location in explaining the growth of settlements. As this was an independent piece of work, Jack demonstrates the ability to follow through a short structured sequence of enquiry. These skills might be developed further when using a range of maps to investigate localities.

(Geographical Enquiry and Skills, Patterns and Processes)

20

Programme of study:
1a, 1b, 1d, 3a, 3e, 4, 5a, 5e

My view of St Lucia

These Year 6 children had already studied the life of a family living in Castries, St Lucia, as part of their Year 3 geography work. The island was now 'revisited' in order to gain an overview of its landscape and character. A class discussion resulted in the identification of some geographical questions. After showing slides of land use and economic activities on St Lucia, the teacher asked the children to work with some reference material about the island. The task was to highlight key words, to classify them using the questions identified and then to present two views of the island – one in words and one in a picture.

In these extracts, Sadie's classification of key words shows her ability to respond to the questions, *eg What makes this place different?*, to identify the meanings of geographical terms and to make the appropriate selection, relevant to St Lucia.

(Geographical Enquiry, Places)

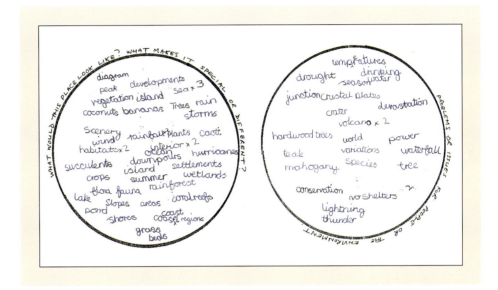

Sadie's picture presents a strong sense of a lush tropical island, but with the dominating presence of volcanic landforms. In this sense, she is highlighting her view of *What makes this place special?* and providing an overview of the St Lucian landscape. This was developed further in the written work which followed.

(Geographical Enquiry, Places)

21

Programme of study:
1a, 1d, 2, 3e, 6, 10

In the news

A Year 6 class was studying the media. The children had already studied interactions between people and their environment and flooding as a hazard, during geographical work on rivers. Discussion was now widened to include a range of natural hazards affecting people and places, because the teacher wanted children to appreciate the wider scale and context of hazardous events. Over a period of two months, the children and teacher together monitored the national newspapers for relevant articles. After class discussion and study of these materials, each child was asked to present the information in a way which highlighted the geographical location and the impact of the hazards.

Sean has produced two different approaches to the task. His map of Britain locates key hazardous events arising from the passage of Hurricane Lili and provides a brief commentary on the impact. His 'In the News' diagram is a simple but effective way to compare the impact of hazards occurring in different countries. These pieces of work and the teacher's comment demonstrate Sean's ability to select relevant information from topical secondary sources and to apply it in a geographical context.

(Geographical Enquiry and Skills, Environmental Relationships and Issues)

Teacher's comment: *Sean was enthusiastic about this project. He used the atlas to find a suitable outline map of Britain for the Hurricane Lili work. He showed initiative by using the TV as a resource for information about the earthquakes in Peru and Cornwall.*

Programme of study:
1b, 1c, 2, 3b, 3f, 8c

What's the weather like?

As part of a longer unit of work on weather, small groups of Year 6 children used weather recording instruments to take weather readings for ten days. They recorded the results individually and also entered the information into a database. The 'fields' were decided by the children. Once the data was entered, the children were shown how to interrogate this and produce graphs. Finally, some questions were set by the teacher and discussed by the whole class. Most children answered the questions individually.

Daniel uses his own weather records and the graphs and daily information he printed out from the computer to make observations about physical trends and processes of change. He provides appropriate explanations for his answers (eg rain from stratus clouds). He has shown that he can undertake a task focused on a specific question and that he can use a range of weather recording instruments and a computer database.

(Geographical Enquiry and Skills, Patterns and Processes)

WEATHER RECORD CHART

Day	mon	Tues	wed
Date	7/10/98	8/10/98	9/10/98
Time	12.45	12.45	12.45
temperature		20°c	19°c
cloud type	stratus		
cloud cover	8oktas		
rainfall			
Beaufort Scale		1	1
wind direction		west	west
wind speed		0	
visibility		Fair	Fair
description			

Date and Temp

Temp

Date

IS OUR WEATHER CHANGEABLE?

The highest temperature we recorded was 20 degrees Celsius. This was on the 8th and 12th October.
The lowest temperature we recorded was 10 degrees Celsius. This was on 9th, 11th and 18thOctober.
The temperature did not stay the same We know this because looking at the highest and lowest temperature they are not the same and if we look at the line on the graph it goes in a different direction<up and down>.

The cloud cover was high .On the days with the lowest temperature thecloud cover was high .
I thought that the temperature and the cloud cover would have a relashonship but it doesn't. I thought that they would have a relashonship because I thought the clouds would act as a blanket.

We did not have much rain.
It was 8 oktas of cloud and the cloud type was stratus
I think there is a relashonship because rain mostly comes from stratus clouds.

A river investigation

After some introductory work about the physical and human geography of rivers, a Year 6 class was taken out to investigate the local river. The work included a river walk, field sketching from the bridge, measurements of river width, depth and speed and some observations of human activities. Much of the work was carried out in groups, although each child had to produce an individual record of the fieldwork activities and their own conclusions. Anna's description and explanation of human activities are not shown here, although they provided further information about her understanding and skills.

Programme of study:
1b, 1c, 2, 3a, 3b, 3c, 5a, 5c, 7b

In these extracts from Anna's fieldwork record, she presents an informative diagram giving the results of a measuring task (river width) and a well-labelled field sketch (based on the teacher's outline, but labelled individually). These provide evidence of her involvement in the fieldwork enquiry tasks and her ability to describe and explain them, including the difficulties of an experiment that didn't work. The diagrams together with the written work (extract only) show that she recognises the physical processes at work and appreciates how they affect the landscape.

(Geographical Enquiry and Skills, Patterns and Processes, Places)

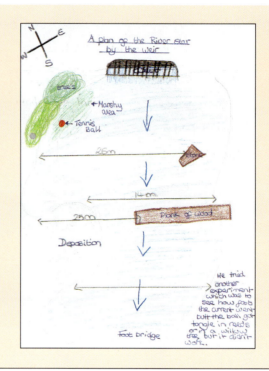

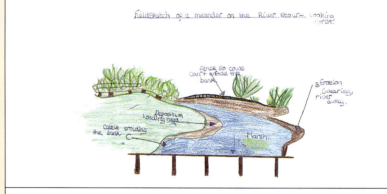